Endangered Species

	Endangered and Threatened Animal Species Worldwide			
	Endangered Species		Threatened Species	
Group	U.S.	Foreign	U.S.	Foreign
Mammals	55	252	9	19
Birds	74	178	16	6
Reptiles	14	65	19	14
Amphibians	7	8	5	1
Fishes	65	11	40	0
Snails	15	1	7	0
Clams	51	2	6	0
Insects	20	4	9	0

[Source: Information Plus — Endangered Species, 1996]

Use the data in the table to answer the question.

Which group contains species that, in number, are:

❶ most endangered in the United States?

❷ most endangered in foreign countries?

❸ most threatened in the United States?

❹ least endangered or threatened worldwide?

❺ more endangered worldwide, fishes or reptiles?

❻ more threatened worldwide, insects or amphibians?

How many are there?

❼ endangered animal species in the U.S.

❽ threatened animal species in the U.S.

❾ mammals that are endangered worldwide

❿ endangered species in foreign countries

⓫ reptiles that are endangered or threatened worldwide

⓬ mammals that are endangered or threatened worldwide

Answer Box

A	B	C	D	E	F
Reptiles	112	335	111	521	301
G	**H**	**I**	**J**	**K**	**L**
307	Fishes	Insects	Mammals	Amphibians	Birds

Objective: Interpret/compare data given in a table.

1

When Do We Leave?

Use this train schedule to answer the question.

Train Schedule From Penn Station, New York City, Monday Through Friday					
Leaves New York	**Millburn**	**Short Hills**	**Summit**	**Chatham**	**Madison**
7:09 A.M.	7:42 A.M.	7:45 A.M.	7:50 A.M.	7:56 A.M.	8:00 A.M.
7:43	8:16	—	8:24	8:30	8:34
8:08	8:40	8:43	8:47	8:52	8:56
9:10	9:42	9:45	9:49	9:54	9:58
10:17	10:52	10:55	10:59	11:04	11:08
11:20	11:52	11:55	11:59	12:04 P.M.	12:08 P.M.

How many minutes does it take to travel:

1. from New York to Madison on the 9:10?
2. to Chatham from NY on the 8:08 train?
3. to Summit from NY on the 7:43 train?
4. from NY to Short Hills on the 10:17 train?
5. to Summit from Millburn on the 11:20 train?
6. from Summit to Chatham on the 7:09 train?

Which is the latest train from New York that a passenger can take to arrive:

7. in Chatham before 10:00 A.M.?
8. in Madison by noon?
9. in Short Hills before 8:30 A.M.?
10. in Summit before noon?
11. in Chatham before 9:00 A.M.?
12. in Millburn by 8:30 A.M.?

Answer Box

A	B	C	D	E	F
6	10:17 A.M.	7:09 A.M.	41	9:10 A.M.	48
G	**H**	**I**	**J**	**K**	**L**
7:43 A.M.	44	7	8:08 A.M.	38	11:20 A.M.

2   Objective: Interpret/compare data given in a schedule.

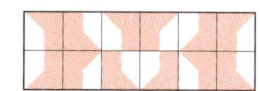

Water Pollution

Fish kills, which mean large groups of fish dying in a lake or other waterway, are often caused by pollution.

How many states reported:

1 1 or more fish kills?

2 over 30 fish kills?

3 more than 70 fish kills?

4 fish kills or did not report?

5 0 fish kills or did not report?

6 fewer than 31 kills but at least 1?

Number of Fish Kills Reported Nationwide in 1990	
Number of Fish Kills	**State Tally**
0 or not reported	‖‖‖ ‖‖‖ ‖
1–10	‖‖‖ ‖‖‖ ‖‖‖ ‖‖‖
11–30	‖‖‖ ‖‖
31–70	‖‖‖ ‖
more than 70	‖‖‖ ‖‖

[Source: Information Plus — Garbage, 1994]

Answer the question.

7 The most reports were made for a range of kills between 1 and 10. True or false?

8 The same number of states reported 11 to 30 kills as 31 to 70 kills. True or false?

9 Can you tell from this table how many states have water pollution problems today? Yes or no?

10 Is it likely that more than half the states in the U.S. had some water pollution in 1990? Yes or no?

11 Jerry says that 7 states reported exactly 30 fish kills. Marsha says that 7 states reported at least 11 kills but no more than 30. Who is correct?

12 Marsha says that the same state may have been included in more than one row under the tally column. Jerry says that each state could only be counted once. Who is correct?

Answer Box

A	B	C	D	E	F
No	12	50	38	13	Jerry
G	**H**	**I**	**J**	**K**	**L**
Marsha	7	True	Yes	False	25

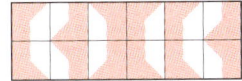

Objective: Interpret/compare data given in a frequency table.

3

Home on the Range

The **range** is the difference between the greatest and the least numbers in a set of data.

Example

Find the range of the set of data.

14, 9, 39, 26, 7

39 − 7 = 32

So, **32** is the range.

Find the range of the set of data.

1. 18, 19, 3, 4, 10, 5
2. 100, 109, 99, 87, 75, 90
3. 34, 29, 44, 10, 5, 50
4. 208, 137, 44, 39, 250, 70
5. 2, 4, 9, 32, 29, 10
6. 49, 55, 55, 80, 41, 30
7. 2, 5, 3, 4, 1, 6
8. 15, 14, 18, 19, 11, 10
9. 34, 20, 44, 50, 20, 100
10. 175, 190, 205, 274, 220, 150

Find the answer.

11. A serving of each of these fresh fruits has the calories shown: apple, 80; apricot, 60; banana, 120; blueberries, 100; cherries, 90; and cantaloupe, 50. What is the range of calories?

12. A serving of the same fruits has the grams of fiber shown: apple, 5; apricot, 1; banana, 3; blueberries, 3; cherries, 3; and cantaloupe, 1. What is the range of grams of fiber?

Answer Box

A	B	C	D	E	F
9	4	5	124	70	30
G	H	I	J	K	L
34	50	211	16	80	45

4 **Objective:** Determine the range for a given set of data.

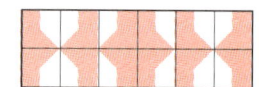

What's My Range?

Find the range.

1. 426, 220, 50
2. $35, $29, $22, $30, $10
3. 4, 5, 6, 2, 3, 4, 9, 9, 6, 7
4. 105, 330, 440, 208
5. $355, $123, $50, $100
6. 10, 20, 30, 40, 10, 20, 30
7. 2.5, 12, 3.9, 1, 15, 12.4, 18
8. 77, 89, 83, 57, 55
9. $993, $912, $409, $290
10. 28, 33.6, 39, 4.1, 22, 3
11. $35, $82, $15, $29, $33
12. $21, $5, $8, $12, $11, $65

Answer Box

A	B	C	D	E	F
335	36	7	$67	30	34
G	H	I	J	K	L
17	376	$305	$703	$60	$25

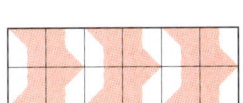

Find the range.

1. $446, $440, $70
2. $37, $49, $44, $30, $30
3. $4, $7, $6, $4, $3, $9, $9
4. $307, $330, $440, $405
5. $377, $343, $70, $300
6. $30, $40, $30, $40, $30
7. $4.70, $34, $3.90, $3, $37
8. $77, $59, $77, $77
9. $993, $934, $409, $490
10. $45, $33.60, $39, $4, $44
11. $37, $54, $37, $49, $33
12. $43, $67, $5, $34, $33, $40

Answer Box

A	B	C	D	E	F
$376	$133	$62	$19	$584	$6
G	H	I	J	K	L
$41	$34	$10	$18	$21	$307

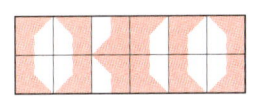

Objective: Determine the range for a given set of data.

What's My Mode?

The **mode** is the number that occurs most often in a set of data.

Find the mode.

1. number of visitors to the zoo each hour:
   149, 148, 145, 145, 145, 144, 105

2. number of miles jogged each week:
   9, 19, 8, 9, 8, 3, 18, 9, 10

3. number of students in each class:
   28, 31, 25, 26, 24, 25, 28, 28, 30

4. math quiz scores:
   10, 9, 8, 8, 9, 10, 10, 9, 10

5. height in inches of players on a team:
   72, 73, 75, 68, 77, 75, 73, 75

6. number of cars passing through a tunnel each hour:
   155, 105, 151, 150, 105

7. admission paid for a movie:
   $4, $4, $4, $4, $4, $4, $2.50

8. money spent on lunch each week:
   $3.50, $4, $2.50, $2.50, $2.50

9. number of minutes spent on school bus:
   20, 25, 18, 28, 30, 19, 31, 30

10. reading test scores:
    88, 90, 89, 94, 93, 99, 90, 75

11. number of days in a month:
    31, 28, 31, 30, 31, 30, 31, 31, 30, 31, 30, 31

12. monthly normal temperatures (in degrees Fahrenheit) for Charleston, South Carolina:
    48, 51, 58, 65, 73, 78, 82, 81, 76, 67, 58, 52

Answer Box

A	B	C	D	E	F
90	$4	105	9	31	75

G	H	I	J	K	L
10	145	58	$2.50	28	30

6 **Objective:** Determine the mode for a given set of data.

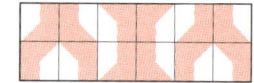

More Math Modes

Find the mode.

1. number of people (in millions) visiting the 10 most popular sites in the national park system during 1996: 17.2; 14.0; 9.4; 9.3; 6.4; 6.1; 6.1; 6.1; 4.9; 4.7

2. number of native speakers (in millions) of the principal languages of the world: 863, 357, 352, 335, 200, 200, 173, 166, 125, 99, 75, 57

3. total number of speakers (in millions) of the principal languages of the world: 1,025; 476; 409; 497; 207; 235; 187; 279; 126; 126; 127; 170

4. areas (in square miles) of notable deserts in the United States: 3,300; 15,000; 70,000

5. number of watts per lightbulb: 60, 90, 40, 20, 100, 60, 150

6. elevations (in feet) of famous waterfalls in Asia: 330, 830, 330

7. number of gold medals won by the top 10 medal winning countries in the 1996 Summer Olympics: 44, 20, 26, 16, 9, 15, 13, 7, 9, 9

8. number of bronze medals won by the top 10 medal winning countries in the 1996 Summer Olympics: 25, 27, 16, 12, 23, 15, 12, 5, 8, 12

9. number of runs batted in by the National League leading players from 1987 to 1997: 137, 109, 125, 122, 117, 109, 123, 116, 128, 150, 140

10. number of home runs hit by the National League home run leaders from 1988 to 1997: 39, 47, 40, 38, 35, 46, 43, 40, 47, 49

11. winning times (in seconds) in the men's 100-m run in the Summer Olympic games held from 1896 to 1936: 12; 11.0; 11.0; 10.8; 10.8; 10.8; 10.6; 10.8; 10.3; 10.3; 10.6

12. distances (in feet) from home plate to center field in National League baseball stadiums: 402, 401, 400, 404, 415, 410, 400, 395, 404, 410, 408, 400, 402, 405, 400, 402

Answer Box

A	B	C	D	E	F
126	109	200	330	No mode	6.1

G	H	I	J	K	L
10.8	60	12	400	40 and 47	9

Objective: Determine the mode for a given set of data.

Explain What You Mean!

Find the mean.

1. pages in newspapers: 25, 16, 39, 45, 11, 12, 10, 10
2. fish in tanks: 12, 3, 4, 22, 9
3. dogs in training: 2, 3, 5, 8, 4, 7, 6
4. baby elephants in African parks: 15, 12, 5, 3, 7, 4, 10
5. paintings in galleries: 45, 23, 16, 22, 7, 8, 33
6. flags in parades: 16, 7, 25, 100, 54, 8
7. marbles in collections: 334, 56, 398, 245, 887
8. points scored in games: 35, 40, 25, 25, 30, 8, 12
9. miles traveled on a bicycle: 258, 331, 247, 328
10. liters of water drunk: 956, 988, 378, 814, 339
11. distances between cities in miles: 458, 552, 296, 773, 1,006
12. diplomas given at graduations: 355, 285, 145, 55, 224, 100

The mean, or the average, of a set of data is the sum of the data divided by the number of items.

Answer Box

A	B	C	D	E	F
194	10	8	21	5	617
G	**H**	**I**	**J**	**K**	**L**
35	25	695	384	22	291

8 Objective: Determine the mean for a given set of data.

By All Means!

Find the mean of the set of data, rounded to the nearest whole number.

1. 73, 78, 82, 97, 89
2. 73, 78, 82, 97, 89, 98
3. 3, 5, 6, 7, 8, 9, 1
4. 3, 5, 6, 7, 8, 9, 8, 9, 7
5. 8, 8, 8, 8, 10, 12, 16, 16
6. 40, 40, 43, 43, 42, 48, 41
7. 10, 10, 12, 12, 23, 34
8. 10, 10, 12, 12, 23, 34, 45
9. 8, 10, 10, 12, 12, 23, 34, 45
10. 40, 40, 43, 43, 42, 48, 41, 54
11. 15, 14, 16, 16, 19, 23, 35
12. 15, 14, 16, 16, 19, 23, 25

Answer Box

A	B	C	D	E	F
11	19	17	44	18	86
G	H	I	J	K	L
6	20	7	84	42	21

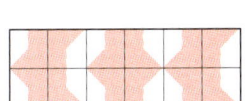

Find the mean of the set of data, rounded to the nearest whole number.

1. 51, 48, 45, 44, 39, 41, 46
2. 4, 3, 13, 19, 26, 30, 35
3. 799, 837, 915, 926, 970
4. 370, 385, 421, 470, 492
5. 311, 313, 312, 328, 342
6. 675, 578, 441, 358, 324
7. 210, 171, 106, 85, 91, 87
8. 81, 72, 68, 63, 63, 65, 51
9. 53, 52, 45, 43, 58, 54, 50
10. 527; 648; 841; 1,028; 1,165
11. 529, 592, 643, 721, 808, 868
12. 704, 731, 734, 714, 718, 725

Answer Box

A	B	C	D	E	F
842	428	51	321	66	475
G	H	I	J	K	L
45	125	694	889	19	721

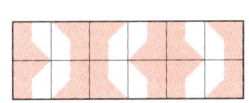

Objective: Determine the mean for a given set of data.

Down the Middle

Find the median.

1. miles driven: 256, 145, 198, 112, 108
2. people attending games: 2,384; 1,224; 2,232; 1,312; 3,211
3. animals photographed: 15, 24, 22, 13, 25, 11, 19
4. daily high temperatures in degrees Fahrenheit: 89, 88, 67, 82, 85, 80, 90
5. monthly rainfall in inches: 2, 3, 0, 1, 1, 1, 2, 2, 1, 1, 3, 0, 1
6. books read: 13, 14, 12, 15, 15, 8, 10
7. hours spent practicing: 2, 3, 1, 3, 4, 1, 1
8. students graduating: 148, 122, 145, 201, 156
9. baby animals born at the zoo: 13, 8, 22, 15, 19, 12, 22
10. hours spent sleeping: 8, 9, 7, 8, 8, 9, 10
11. pounds of vegetables harvested: 79, 85, 67, 60, 58
12. miles flown: 2,384; 3,133; 4,580; 2,178; 1,354; 4,379; 1,933

The median is the middle number when the data are arranged in order from least to greatest.

VISIT THE PETTING ZOO'S NEW ARRIVALS!!!!!

Answer Box

A	B	C	D	E	F
2,232	13	15	19	145	85

G	H	I	J	K	L
148	2,384	67	1	8	2

Objective: Determine the median for a given set of data.

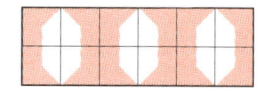

Median Strip

Find the median.

1) 12, 14, 11, 37, 39, 10
2) 12, 18, 25, 13, 23, 11
3) 126, 225, 102, 221, 142, 166
4) 3, 2, 1.5, 1, 1, 2.5
5) 122, 114, 67, 89, 91, 113
6) 487, 239, 107, 96, 108, 225
7) 98, 78, 99, 75, 88, 99
8) 168, 33, 45, 139, 133, 28
9) 487, 525, 388, 184, 287, 167
10) 14, 18, 23, 19, 35, 29
11) 2, 3, 3, 4, 1, 1, 0, 0, 0, 5, 5, 5
12) 345, 268, 407, 311, 614, 552

Answer Box

A	B	C	D	E	F
337.5	102	1.75	13	89	166.5
G	H	I	J	K	L
2.5	154	93	21	15.5	376

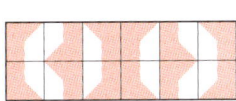

Find the median.

1) 160, 165, 300, 160, 600
2) 36, 41, 44, 79, 78, 78
3) 6, 10, 7, 9, 8, 8
4) 10, 16, 12, 14, 12, 13
5) 2, 8, 2, 7, 6, 5, 2, 5
6) 36, 79, 41, 78, 42, 78, 44
7) 9, 16, 10, 14, 12, 13, 12
8) 2, 9, 2, 8, 2, 7, 8, 5, 6, 5
9) 0, 6, 1, 2, 1, 2, 5
10) 10, 16, 12, 16, 12, 14, 13
11) 150, 900, 160, 600, 160, 300
12) 150, 155, 600, 160, 300, 160

Answer Box

A	B	C	D	E	F
230	8	61	44	165	12
G	H	I	J	K	L
2	5	160	5.5	13	12.5

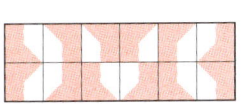

Objective: Determine the median for a given set of data.

Fabulous Fruit!

Use the line plot to answer the question.

How many grams of fiber are in:

1. a serving of fresh oranges?
2. a serving of fresh pears?
3. 2 servings of watermelon?
4. an apple and a banana?

Grams of Fiber per Serving of Fresh Fruit

	Apricots	Apple	Banana	Grapefruit	Grapes	Orange	Pear	Peach	Raspberries	Watermelon	Cherries	Tangerine	Plum
8									×				
7									×				
6				×					×				
5		×		×		×			×				
4		×		×		×	×		×				
3		×	×	×		×	×		×		×		
2		×	×	×	×	×	×		×		×	×	
1	×	×	×	×	×	×	×	×	×	×	×	×	×

[Source: Information Plus — Nutrition, 1995]

Which fruit has:

5. the most fiber per serving?
6. 6 grams of fiber per serving?
7. half as much fiber as raspberries per serving?
8. as much fiber per serving as an orange?
9. the same amount of fiber as cherries per serving?
10. as much fiber per serving as a tangerine?
11. more fiber, a serving of cherries or a serving of grapes?
12. more fiber per serving, an orange or apricots?

Answer Box

A	B	C	D	E	F
Banana	2	4	Grapefruit	8	5
G	**H**	**I**	**J**	**K**	**L**
Raspberries	Cherries	Apple	Orange	Grapes	Pear

12 **Objective:** Interpret/compare data given in a line plot.

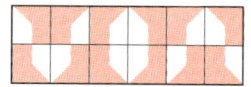

Let's Read!

Use the line plot to answer the question.
Who read:

① the most books?

② half as many books as Rico?

③ the least number of books?

④ the same number of books as Paul?

Answer the question.

⑤ Rosa and Alana together read more books than Rico did. True or false?

⑥ What is the range of books read?

⑦ Did Paul read more or fewer books than Rosa?

⑧ How many books did Julia read?

⑨ How many books did Chester read?

⑩ Did Alana read more or fewer books than Rosa?

⑪ Bo and Alana together read the same number of books that Hiroshi read. True or false?

⑫ What is the mean number of books read, to the nearest whole number?

Books Read

	Bo	Hiroshi	Alana	Julia	Rosa	Chester	Rico	Paul

Answer Box

A	B	C	D	E	F
6	Chester	Bo	5	More	Rico
G	**H**	**I**	**J**	**K**	**L**
7	9	True	False	Hiroshi	Fewer

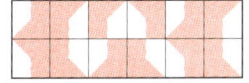

Objective: Interpret/compare data given in a line plot.

13

Underwater Tunnels

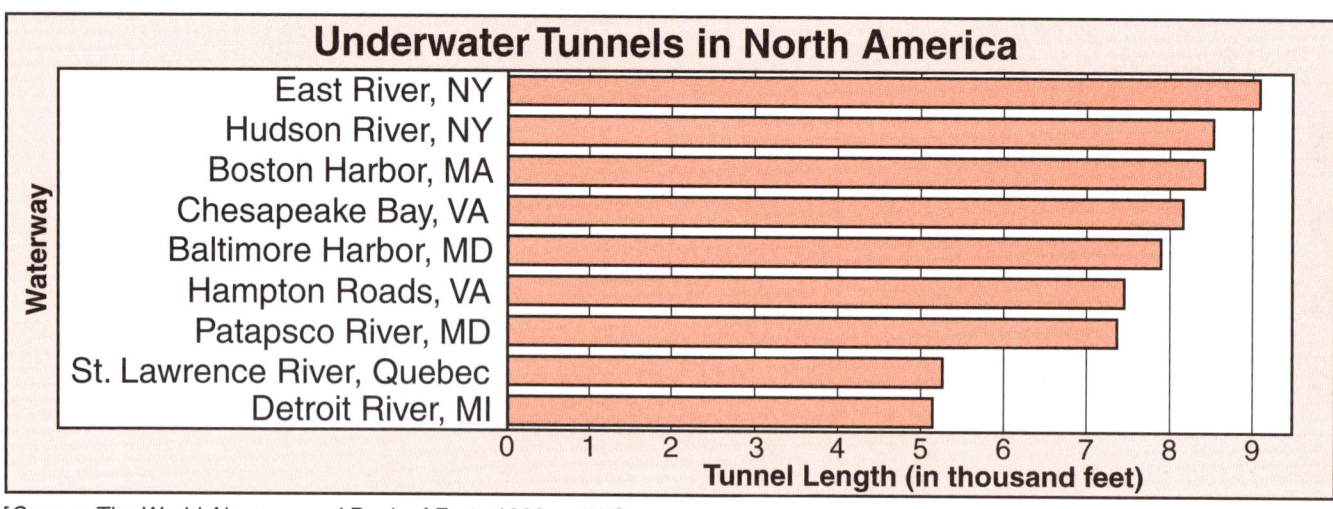

[Source: The World Almanac and Book of Facts 1998, p. 712]

Use the graph to answer the question.

1 The Baltimore Harbor tunnel is more than twice as long as the Detroit River tunnel. Yes or no?

2 The length of more than half the tunnels is at least 7,000 ft. True or false?

3 Is the shortest tunnel on the graph more than or less than 6,000 ft long?

4 The East River tunnel is about 500 ft longer than the Hudson River tunnel. Yes or no?

5 Is the mean more than or less than the length of the Detroit River tunnel?

6 The East River tunnel is 6,000 ft long. True or false?

To the nearest thousand feet, what is the length of the:

7 longest tunnel?

8 Boston Harbor tunnel?

9 Detroit River tunnel?

10 Patapsco River tunnel?

11 range of the tunnels?

12 difference between the two Virginia tunnels?

Answer Box

A	B	C	D	E	F
4,000	More than	9,000	No	8,000	1,000

G	H	I	J	K	L
5,000	Less than	Yes	7,000	True	False

14 Objective: Interpret/compare data given in a horizontal bar graph.

Skyscrapers

Which building on the graph has:

1. the most stories?
2. about half as many stories as the Hancock Tower?
3. 47 stories?
4. between 50 and 55 stories?
5. 37 stories?
6. the same number of stories as Exchange Place?

For the number of stories in the buildings, what is the:

7. mode?
8. range?
9. mean?

How many buildings have:

10. fewer than 50 stories?
11. 50 or more stories?
12. between 40 and 50 stories?

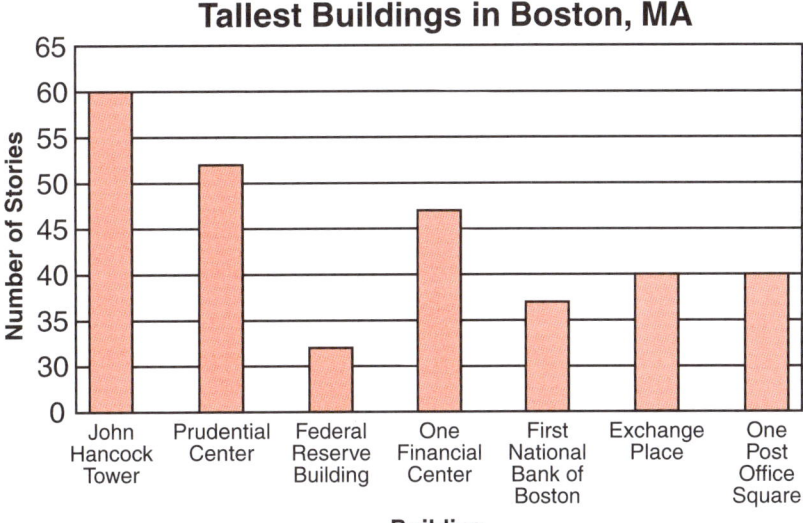

This bar graph shows how many stories tall some of the tallest buildings in Boston are.

Answer Box

A	B	C	D	E	F
Federal Reserve Building	One Post Office Square	5	40	One Financial Center	44
G	H	I	J	K	L
1	28	John Hancock Tower	Prudential Center	2	First National Bank of Boston

Objective: Interpret/compare data given in a vertical bar graph.

Problem Solving: Using a Diagram

Use the diagram to solve the problem.

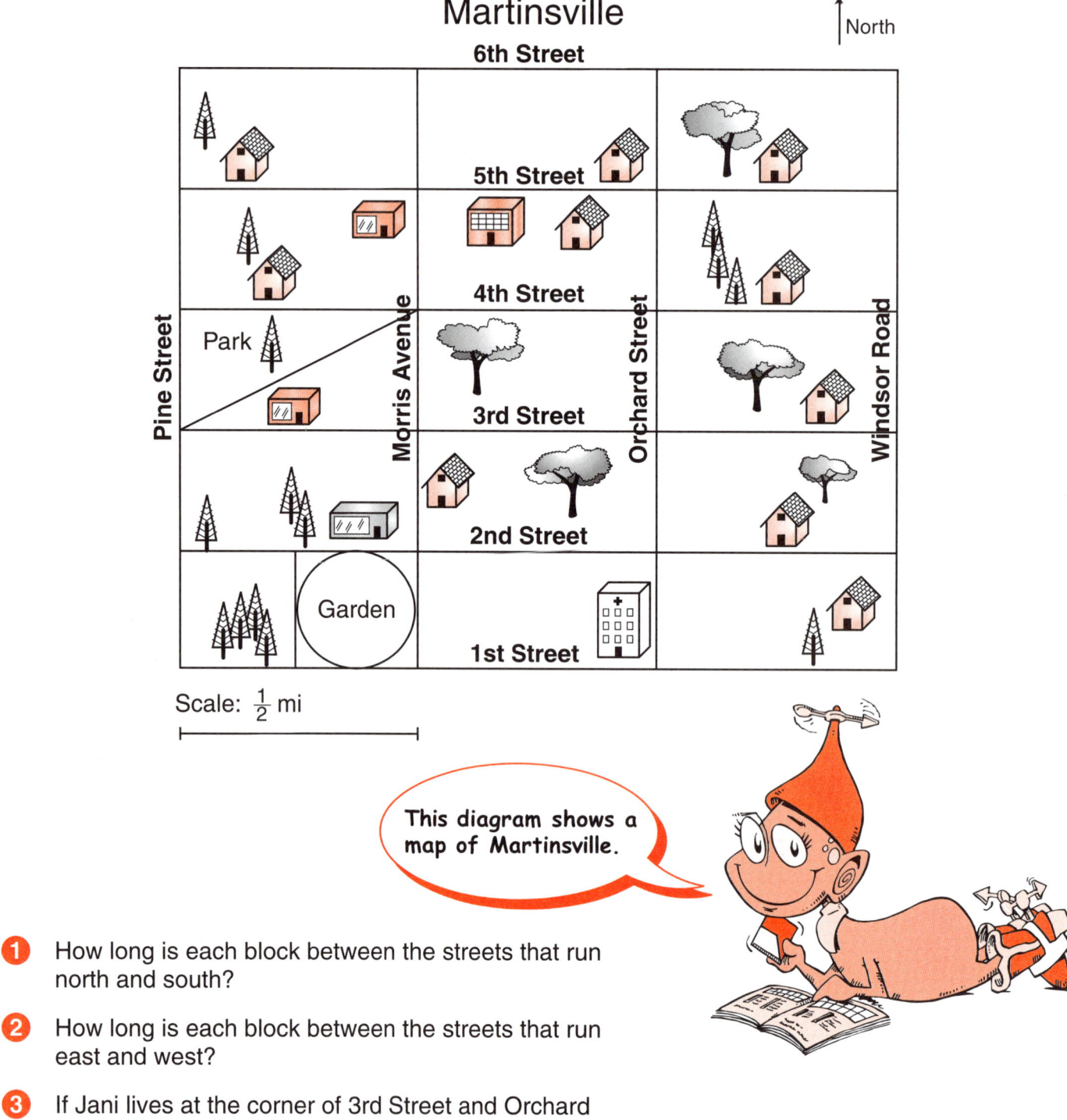

1. How long is each block between the streets that run north and south?

2. How long is each block between the streets that run east and west?

3. If Jani lives at the corner of 3rd Street and Orchard Street, how far must she bike to reach her friend's house at the corner of 5th Street and Morris Avenue?

④ There are 3 different ways Jani can get to her friend's house and each is the same number of miles long. True or false?

⑤ How far must Jani travel from her house to reach the post office at the corner of 6th Street and Orchard Street?

⑥ The distance from Pine Street to Windsor Road is the same distance as from 1st Street to 5th Street. True or false?

⑦ What is the area of each block in Martinsville?

⑧ What is the perimeter of each block?

⑨ One of the blocks between Pine Street and Morris Avenue has a fence along a diagonal that divides the block into 2 parts. One part is a park. How many square miles does the park cover?

⑩ Another block between Pine Street and Morris Avenue has been used for a community garden. The fence around the garden is circular, and the diameter of the garden extends from 2nd Street to 1st Street. What is the radius of the garden?

⑪ Does Bessie bicycle more than 3 mi, if she starts at Morris Avenue and 1st Street and bicycles up Morris Avenue to 6th Street and then across 6th Street to Windsor Road? Yes or no?

⑫ Each house along 5th Street must be on a lot at least 100 ft wide. Is it possible for 26 houses to be on one side of 5th Street between Morris Avenue and Orchard Street? Yes or no?

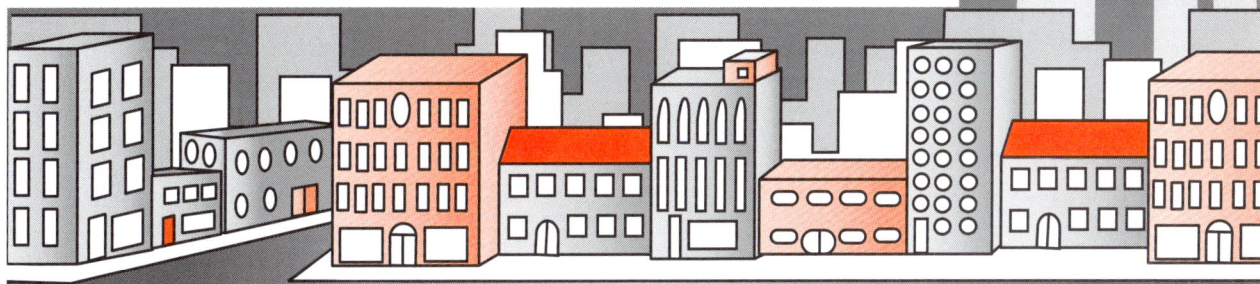

Answer Box

A	B	C	D	E	F
$\frac{1}{8}$ mi	$\frac{1}{8}$ mi²	$\frac{1}{16}$ mi²	1 mi	Yes	$1\frac{1}{2}$ mi
G	**H**	**I**	**J**	**K**	**L**
True	$\frac{1}{4}$ mi	$\frac{3}{4}$ mi	No	$\frac{1}{2}$ mi	False

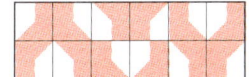

Objective: Solve a problem by using information in a diagram.

Look at All the People!

Use the line graph to answer the question.

To the nearest billion, what was the world population in:

1 1997?

2 1960?

3 1990?

4 1980?

In what 10-year period did the world population increase:

5 by about half a billion people?
1950–1960 or 1980–1990

6 to over 4 billion?
1960–1970 or 1970–1980

7 to over 3 billion?
1950–1960 or 1960–1970

8 at a faster rate?
1950–1960 or 1980–1990

Answer the question.

9 Is it likely that the population will be greater than 6 billion by 2020? Yes or no?

10 The world population in 1950 was about half of what it was in 1985. True or false?

11 Is it likely that the population was less than 1 billion in 1945? Yes or no?

12 The population of the world more than tripled from 1950 to 1997. True or false?

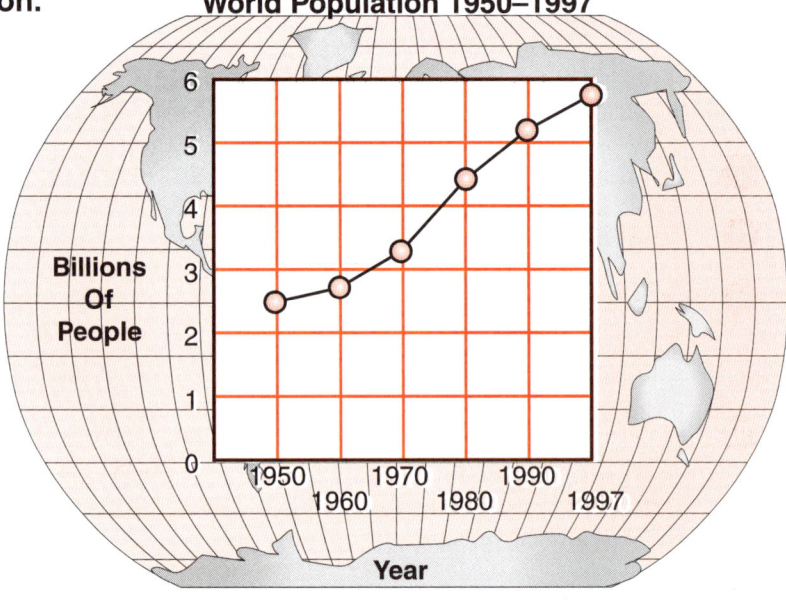

[Sources: World Almanac 1998, p. 838 and Information Plus–Endangered Species, 1994, p. 9]

Answer Box

A	B	C	D	E	F
1960–1970	6 billion	False	3 billion	1950–1960	Yes
G	**H**	**I**	**J**	**K**	**L**
4 billion	True	1970–1980	1980–1990	5 billion	No

18 Objective: Interpret/compare data given in a line graph.

Save Your Money!

Use the graph to answer the question.

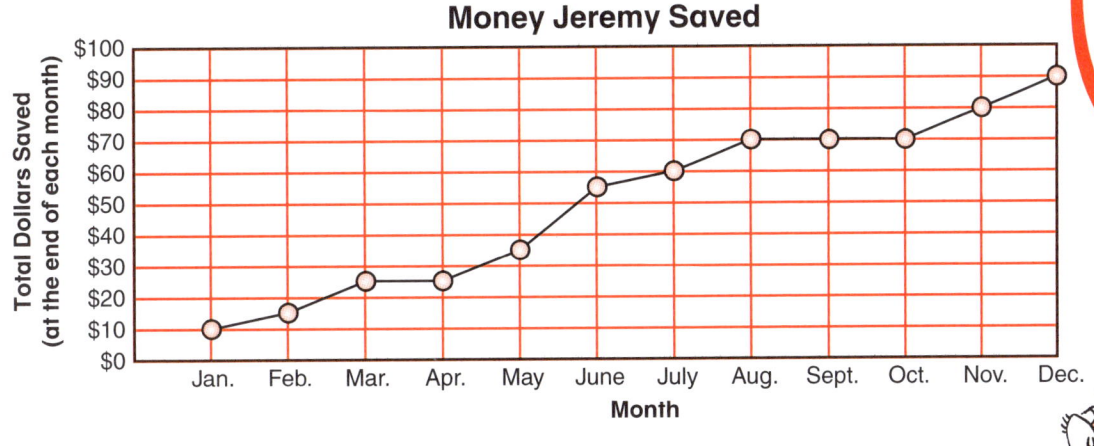

How much money did Jeremy save:

1. by the end of July?
2. by the end of February?
3. by the end of June?
4. from July to December?
5. in February?
6. in September?

Answer the question.

7. Did he save more money in February or April?
8. What is the most Jeremy saved in any one month?
9. What is the mode of his monthly savings
10. Which was the first month in which he saved no money?
11. Jeremy did not save any money in 3 out of 12 months. True or false?
12. On average, Jeremy saved about $50 each month. True or false?

Answer Box

A	B	C	D	E	F
$10	$55	False	$20	$60	$30
G	**H**	**I**	**J**	**K**	**L**
$5	True	$0	$15	February	April

Objective: Interpret/compare data given in a line graph.

19

Spend Your Money!

Use the graphs to answer the question.

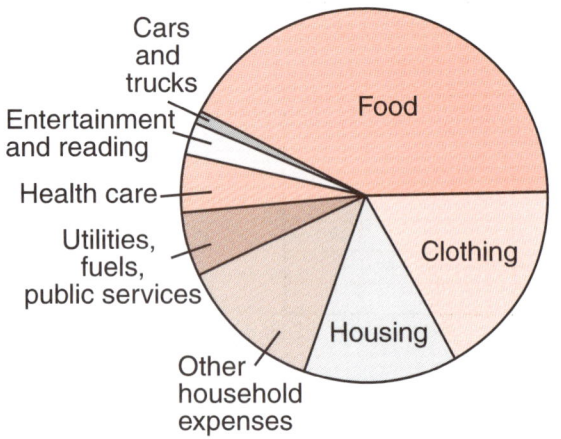

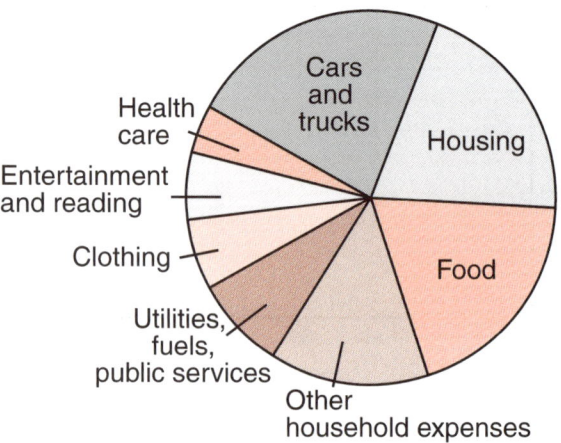

[Source: Information Plus – Nutrition, 1995, p. 136]

Which category of a consumer's budget represents:

1. the largest part in 1917?
2. the largest part 1988?
3. the smallest part in 1988?
4. about 3 times more in 1917 than in 1988?
5. about the same part in 1988 as clothing?
6. about the same part in 1917 as health care?

If the Jones family budget for the items shown was $50,000, about how much would they have spent on:

7. food in 1988? $20,000 or $10,000
8. cars and trucks in 1917? $500 or $12,000
9. clothing in 1988? $3,000 or $9,000
10. food in 1917? $20,000 or $10,000
11. housing in 1917? $3,000 or $9,000
12. cars and trucks in 1988? $12,000 or $500

Answer Box

A	B	C	D	E	F
Utilities	Cars and trucks	$3,000	Entertainment	Clothing	Food
G	H	I	J	K	L
$12,000	$500	$9,000	Health care	$10,000	$20,000

20 **Objective:** Interpret/compare data given in a circle graph.

Borrow It!

Use the graphs to answer the question.

Books Borrowed in May

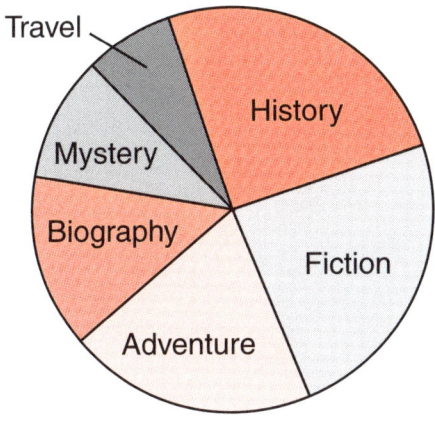

Books Borrowed in June

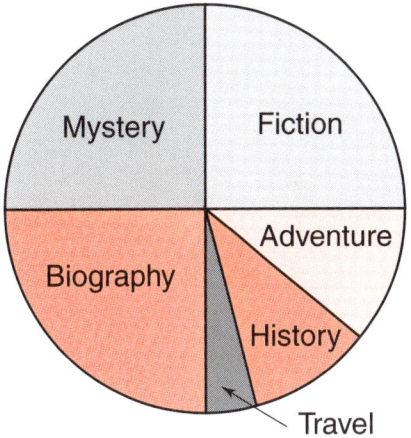

Which type of book:

1. was the least popular in June?
2. was as popular as history in June?
3. was as popular in June as in May?
4. was less popular in June than in May, history or fiction?
5. was more popular in June than in May, mystery or adventure?
6. was less popular in May than in June, travel or biography?

If 1,000 books were borrowed each month from the six categories in the graph, then about how many were:

7. mystery books in June, 250 or 500?
8. adventure books in May, 200 or 120?
9. mystery or fiction books in May, 350 or 650?
10. biography or fiction in June, 250 or 500?
11. adventure books in June, 120 or 200?
12. not mystery or fiction books in May, 350 or 650?

Answer Box

A	B	C	D	E	F
Fiction	350	500	250	Adventure	Biography
G	**H**	**I**	**J**	**K**	**L**
120	Mystery	History	Travel	650	200

Objective: Interpret/compare data given in a circle graph.

21

Problem Solving: Using a Bar Graph

Use the graph to answer the question.

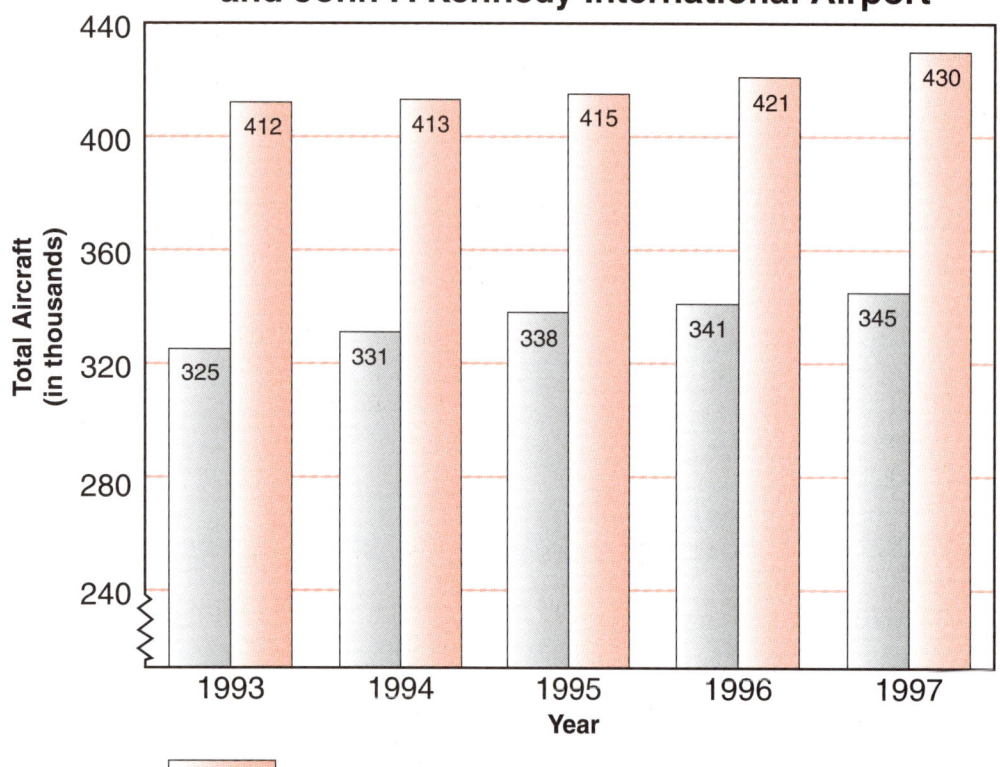

① What is the range of years shown in the graph?

② What is the range for the number of aircraft shown for Newark International Airport?

③ What is the range for the number of aircraft shown for John F. Kennedy International Airport?

④ Knowing the size as well as the number of aircraft using an airport is necessary to predict how many passengers the airport can service. Correct or incorrect?

⑤ The number of aircraft using John F. Kennedy International Airport decreased from 1993 to 1996. True or false?

This **double bar graph** compares the number of aircraft using two major airports.

6 By how much did the total aircraft for both airports increase from 1993 to 1997?

7 Was the difference in the two totals for 1993 greater than the difference in the two totals for 1996? Yes or no?

8 What was the total number of aircraft using both airports in 1994?

9 Newark International Airport served fewer aircraft than John F. Kennedy International Airport from 1993 to 1997. Correct or incorrect?

10 The number of aircraft using Newark International Airport increased the most from 1996 to 1997. True or false?

11 What was the total number of aircraft using both airports in 1997?

12 Was the difference in total aircraft for the two airports ever greater than 100,000 for any single year? Yes or no?

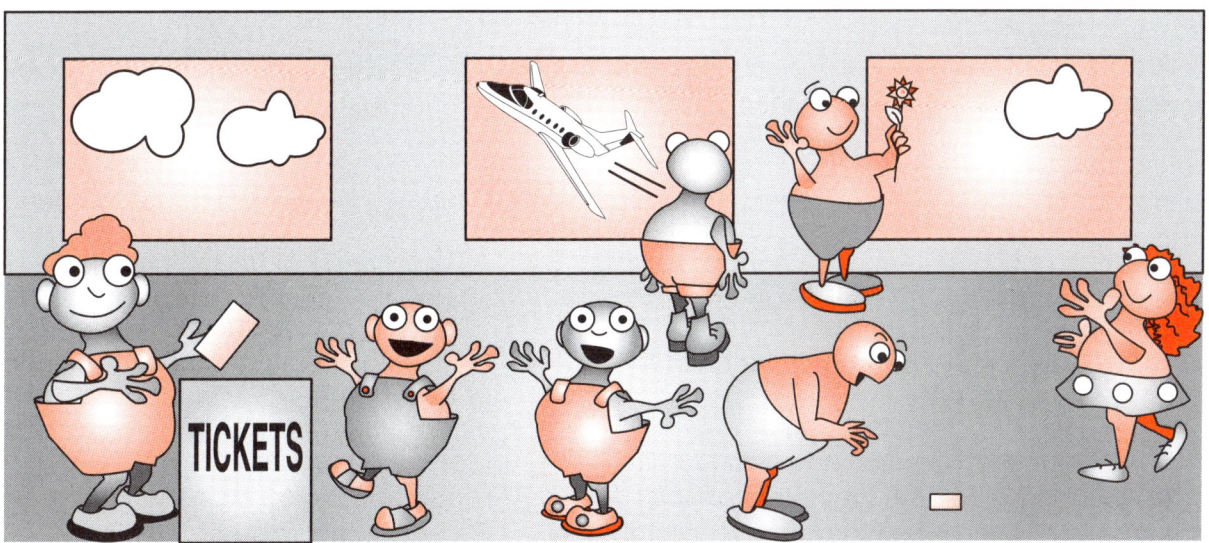

Answer Box

A	B	C	D	E	F
4	Correct	False	775,000	18,000	38,000
G	**H**	**I**	**J**	**K**	**L**
True	Yes	Incorrect	20,000	744,000	No

Objective: Solve a problem by drawing conclusions from data given in a bar graph.

Blow, Wind, Blow!

Use the plot to answer the question.

Highest Wind Speeds (mph)

4	4	6	8		
5	2	3	4	4	8
6	0	0			
7	3	5	6		
8	0	1			
9	9				

4|6 means 46 mph.

[Source: World Almanac 1998, p. 184]

This **stem-and-leaf plot** shows the highest wind speed in miles per hour for 16 cities in the United States.

1 What is the range of the wind speeds?

2 What is the median of the wind speeds?

3 Are there any items of data included that are less than 91 mph and greater than 81 mph? Yes or no?

4 What is the mean of the wind speeds, to the nearest mile per hour?

5 How many items of data are included?

6 What is the mode of the wind speeds?

7 There is only one item of data included in the range of 60 to 69 mph. True or false?

8 A wind speed of 35 mph would be plotted next to the stem 3. Correct or incorrect?

9 Is a wind speed of 54 mph included in the data? Yes or no?

10 How many items of data are included next to the stem 5?

11 A wind speed of 35 mph and a wind speed of 55 mph would both be plotted on the same line of the plot. Correct or incorrect?

12 More items of data are included in the range of 50 to 59 mph than in the range of 70 to 79 mph. True or false?

Answer Box

A	B	C	D	E	F
54 and 60	Correct	5	False	No	Incorrect
G	**H**	**I**	**J**	**K**	**L**
True	59	55	63	16	Yes

24 Objective: Interpret/compare data given in a stem-and-leaf-plot.

What's the Outcome?

Find the missing possible outcome for the spinner.

A **possible outcome** is any result that could occur in a probability experiment.

1
1, 2, ■, 3

2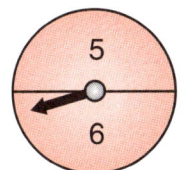
5, ■

3
■, 10, 5, 8, 7, 6

4
■, 8, 10

5
■

6
10, ■, 11

7
■, 4

8
■, 10, 11

9
3, ■, 7, 5

10
10, 6, ■

11
2, ■, 4, 3

12
11, ■

Answer Box

A	B	C	D	E	F
10	7	2	1	3	8

G	H	I	J	K	L
6	12	9	5	4	11

Objective: Complete a list of all the possible outcomes generated by spinning a spinner.

25

Problem Solving: Solving Multi-Step Problems

Solve the problem.

① A shipment of 45 goldfish and 33 tropical fish arrives. If the fish are priced at $1.25 per goldfish and $3.45 per tropical fish, then how much will the shop have taken in when all the fish are sold?

② The fish are separated into smaller tanks. One size tank holds 15 goldfish and another holds 11 tropical fish. How many tanks are needed to hold all the fish in problem 1?

③ One customer orders 25 lb of bird seed and a 50-lb bag of dog food. The bird seed costs $6.50 and the dog food costs $7.45. If dog food can also be purchased for $0.25 per lb, then which is the better buy, by the pound or by the bag?

④ How much does the bird seed cost per pound?

⑤ A customer buys 2 lb of cat snacks for $6.40. How much is this per ounce?

6 A dog leash that measures over 7 ft long will be at least 117 in. long. True or false?

7 A customer spends $22.25 for dog food, not including tax. Is it likely he purchased the food by the bag or by the pound?

8 There are 21 cans in the first row of a display, 15 in the second row, 10 in the third row, and 6 in the fourth row. If the pattern continues, then how many more cans are needed to make the remaining two rows?

9 If the shop is open from 9 A.M. to 5:45 P.M. each weekday and from 8:30 A.M. to 5:00 P.M. on Saturday, then the store is open about 52 hours per week. True or false?

10 A stock clerk earns $5.50 per hour. How much will she earn if she works 2 h every weekday and 8.5 h on Saturday?

11 A box of dog biscuits has 12 beef flavor biscuits, 10 liver flavor biscuits, 6 cheese flavor biscuits, and 2 chicken flavor biscuits. What is the probability that if a pet owner takes a biscuit out of the box without looking, it will be cheese flavor?

12 What is the probability that the biscuit will not be liver flavor?

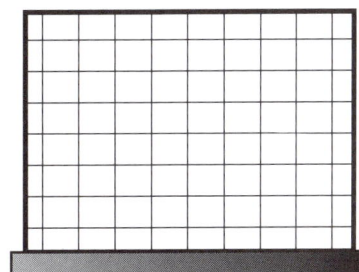

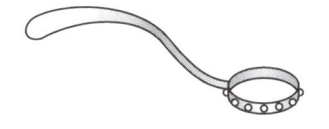

Answer Box

A	B	C	D	E	F
False	6	$0.20	$\frac{1}{5}$	$170.10	$0.26
G	H	I	J	K	L
4	$\frac{2}{3}$	True	By the bag	By the pound	$101.75

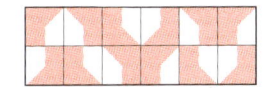

Objective: Solve a multi-step problem involving all operations with whole numbers, decimals, or fractions.

27

Fun with Spinning!

CONSIDER THIS

The **probability** of an event is the number of successful outcomes divided by the number of possible outcomes.

There is a $\frac{2}{5}$ chance of spinning a D.

The probability is $\frac{2}{5}$.

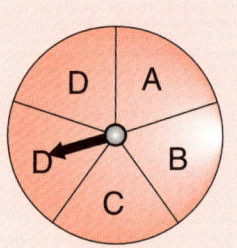

Find the probability.

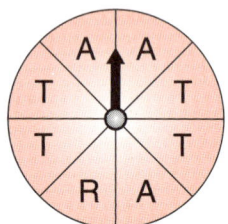

1. spinning the letter A
2. spinning the letter T
3. spinning the letter R

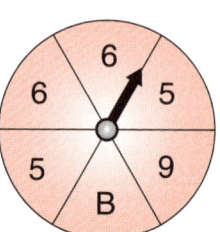

4. spinning the number 6
5. spinning the number 9
6. spinning a number

7. spinning a number greater than 4
8. spinning an even number
9. spinning an odd number

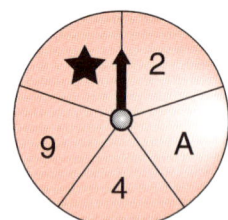

10. spinning a number less than 7
11. spinning the number 2
12. spinning a number or a letter

Answer Box

A	B	C	D	E	F
$\frac{1}{4}$	$\frac{1}{8}$	$\frac{5}{8}$	$\frac{2}{5}$	$\frac{1}{2}$	$\frac{5}{6}$
G	H	I	J	K	L
$\frac{1}{6}$	$\frac{1}{5}$	$\frac{3}{8}$	$\frac{1}{3}$	$\frac{4}{5}$	$\frac{3}{4}$

Objective: Use a fraction to describe the probability of obtaining an outcome when spinning a spinner.

Pick a Card... Any Card!

Find the probability.

Remember! Sometimes the probability of an event is 0 or 1.

1. drawing a white card
2. drawing a striped card
3. drawing a polka-dotted card
4. drawing a black card

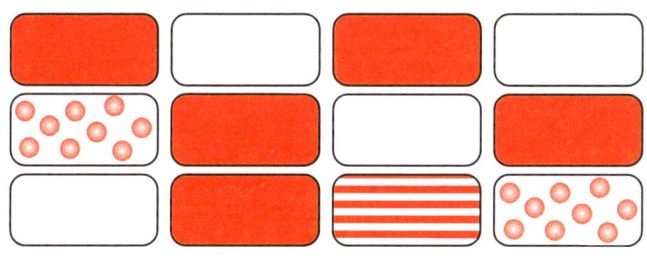

5. drawing a white or orange card
6. drawing an orange, white, striped, or polka-dotted card
7. drawing an orange card
8. drawing a white card

9. drawing a white card
10. drawing an orange card
11. drawing a white or an orange card
12. drawing a card that is not white

Answer Box

A	B	C	D	E	F
$\frac{1}{3}$	$\frac{4}{5}$	$\frac{2}{5}$	$\frac{5}{12}$	$\frac{3}{5}$	$\frac{3}{4}$

G	H	I	J	K	L
$\frac{1}{2}$	1	$\frac{3}{8}$	0	$\frac{1}{5}$	$\frac{1}{8}$

Objective: Use a fraction to describe probability of obtaining an outcome when picking a card from a group.

29

Roll the Cube!

A **number cube** has numbers 1 to 6 on its faces.

Find the probability.

1. rolling a 1
2. rolling a 1 or 2
3. rolling a number
4. rolling a letter
5. rolling an even number
6. rolling a number less than 5

Find the answer.

7. The probability of rolling a 2 on a number cube is not equal to the probability of landing on 2 on the spinner at the right. Correct or incorrect?

8. The probability of an event is never less than 0. True or false?

9. Is the probability of rolling an even number on a number cube equal or not equal to the probability of tossing a head on a penny?

10. The probability of rolling a number less than 6 is greater than the probability of rolling a number greater than 1. Correct or incorrect?

11. The probability of an event is always less than 1. True or false?

12. Is the probability of rolling a number on a number cube equal or not equal to the probability of tossing a tail on a coin?

Answer Box

A	B	C	D	E	F
Correct	$\frac{1}{6}$	Equal	1	$\frac{2}{3}$	True
G	H	I	J	K	L
0	Incorrect	False	$\frac{1}{2}$	Not equal	$\frac{1}{3}$

Objective: Use a fraction to describe the probability of getting an outcome when tossing a number cube.

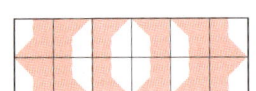

Make a Prediction!

**Predict the number of times the outcome will occur.
The spinner at the right is spun 40 times.**

1. The spinner lands on a letter or polygon.

2. The spinner lands on a triangle or parallelogram.

3. The spinner lands on a letter.

4. The spinner lands on a geometric figure.

5. The spinner lands on a geometric solid.

6. The spinner lands on A, B, C, or R.

7. The spinner lands on a letter or a geometric shape or figure.

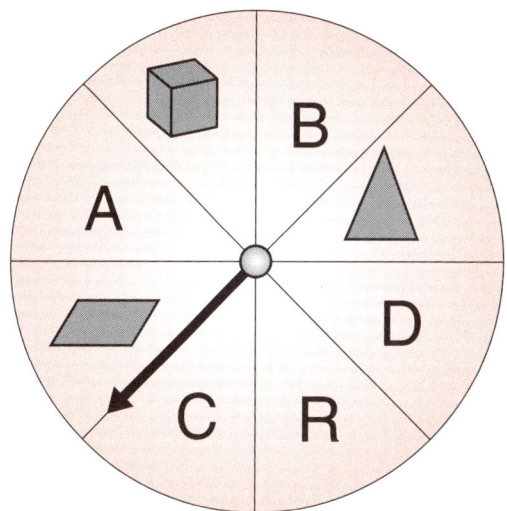

The spinner at the right is spun 20 times.

8. The spinner lands on a star.

9. The spinner lands on a number or letter.

10. The spinner lands on a letter.

11. The spinner lands on a number or star.

12. The spinner does not land on a number, letter, or star.

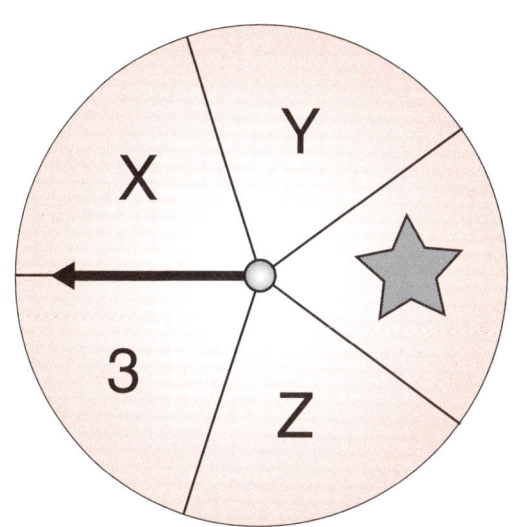

Answer Box

A	B	C	D	E	F
40	12	0	4	5	8
G	H	I	J	K	L
15	35	10	20	25	16

Objective: Given a spinner, predict the number of times an outcome will occur in a given number of spins.

31

More Predictions!

Predict the number of times the outcome will occur.
A cube numbered 1–6 is rolled 30 times.

1. The number 1 is rolled.
2. A number greater than 6 is rolled.
3. The number 2 or 3 is rolled.
4. A number less than or equal to 5 is rolled.
5. A number greater than 2 is rolled.
6. An even number is rolled.

The number cube is rolled 180 times.

7. The number 4, 5, 6, or 3 is rolled.
8. The number 1, 2, 3, 4, 5, or 6 is rolled.
9. An odd number is rolled.
10. The number 4 is rolled.
11. The number 5, 6, or 7 is rolled.
12. A number greater than 1 is rolled.

Answer Box

A	B	C	D	E	F
10	90	150	0	25	5
G	H	I	J	K	L
20	60	15	180	120	30

Objective: Given a number cube, predict the number of times an outcome will occur in a given number of tosses.

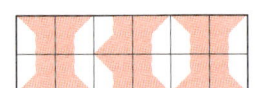